Written and illustrated by Nicholas Senior
Published by Nicholas Senior
Edited by Ellen Wettersten

ISBN: 978-1-7647075-0-3

First edition, 2026

How the 2-Digit Numbers Got Their Names

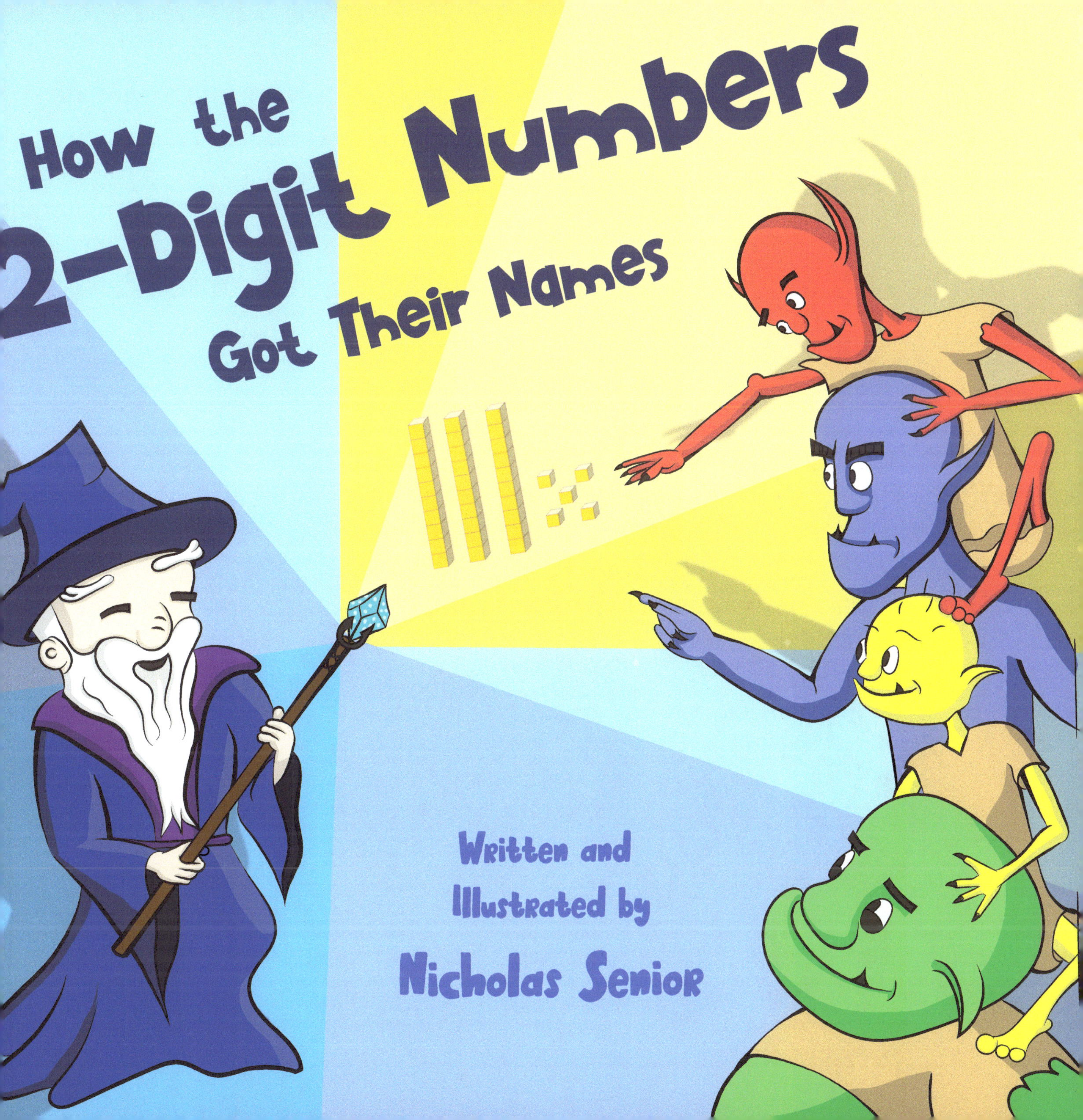

Written and Illustrated by

Nicholas Senior

A long, long time ago, Master of Order had a brilliant idea. He had found a way to make it easy to count large numbers by grouping them into tens.

All he had to do was say how many tens and how many more.

1
2
3
4
5
6
7
8
9
1 ten
1 ten 1
1 ten 2
1 ten 3
1 ten 4
1 ten 5
1 ten 6
1 ten 7
1 ten 8
1 ten 9
2 tens
2 tens 1
2 tens 2
2 tens 3
2 tens 4
2 tens 5
2 tens 6
2 tens 7
2 tens 8
2 tens 9
3 tens
3 tens 1
3 tens 2
3 tens 3
3 tens 4
3 tens 5
3 tens 6
3 tens 7
3 tens 8
3 tens 9
4 tens
4 tens 1
4 tens 2
4 tens 3
4 tens 4
4 tens 5
4 tens 6
4 tens 7
4 tens 8
4 tens 9
5 tens
5 tens 1
5 tens 2
5 tens 3
5 tens 4
5 tens 5
5 tens 6
5 tens 7
5 tens 8
5 tens 9
6 tens
6 tens 1
6 tens 2
6 tens 3
6 tens 4
6 tens 5
6 tens 6
6 tens 7
6 tens 8
6 tens 9
7 tens
7 tens 1
7 tens 2
7 tens 3
7 tens 4
7 tens 5
7 tens 6
7 tens 7
7 tens 8
7 tens 9
8 tens
8 tens 1
8 tens 2
8 tens 3
8 tens 4
8 tens 5
8 tens 6
8 tens 7
8 tens 8
8 tens 9
9 tens
9 tens 1
9 tens 2
9 tens 3
9 tens 4
9 tens 5
9 tens 6
9 tens 7
9 tens 8
9 tens 9

Excited, Master of Order wrote down all the number names, carefully following the simple pattern of counting how many tens and how many more.

He was so happy with these new number names that he decided to enter them into the Super Idea Spreader! Soon, everyone would be able to count large numbers.

Pleased with his hard work, he decided to go and make a cup of tea while the Super Idea Spreader warmed up. But, just as his idea was about to go through, something happened that Master of Order did not expect!

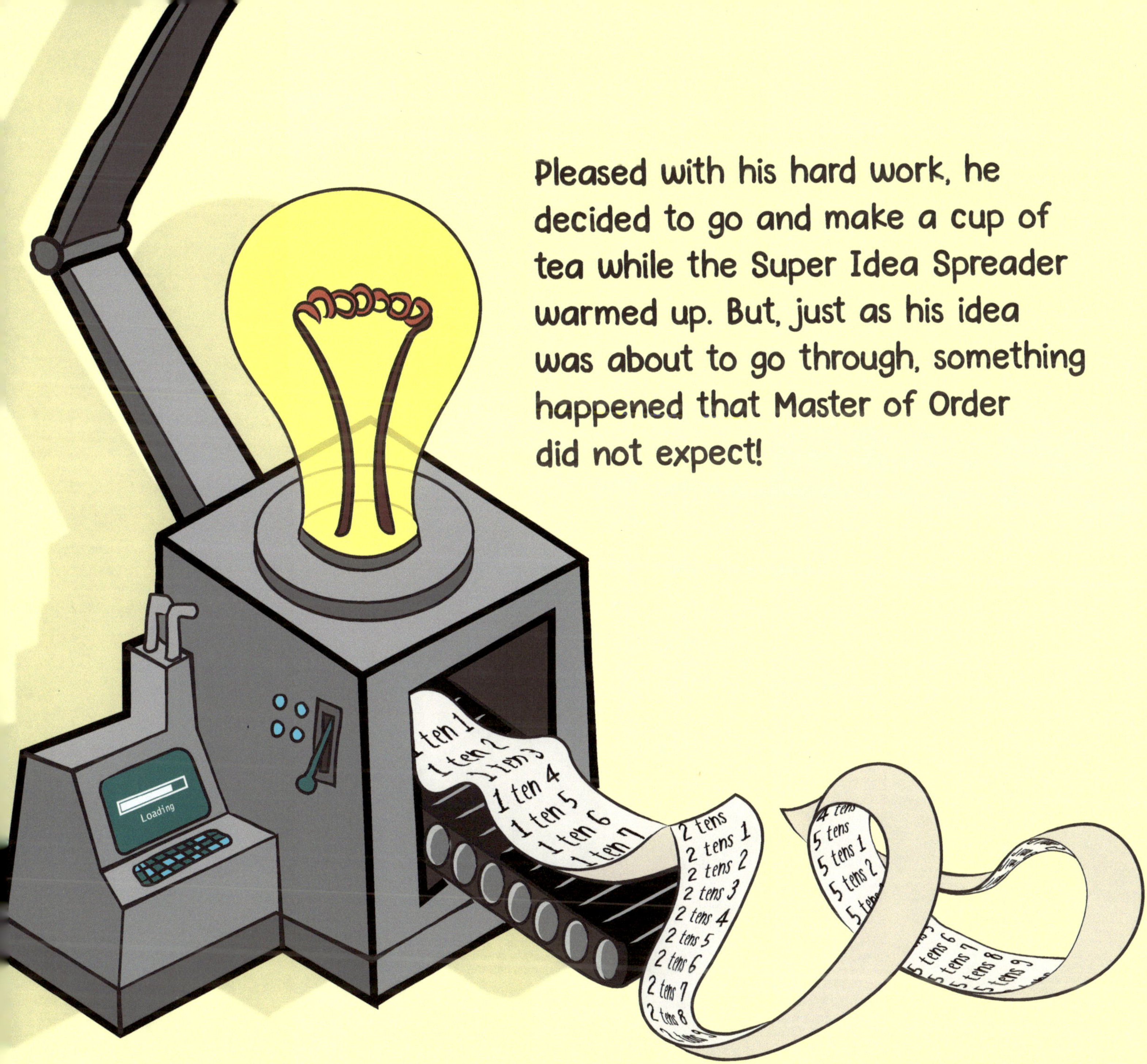

The Goblins of Chaos had pulled the plug at the very last minute and were up to no good.

They pulled the list of numbers out of the Super Idea Spreader, looking to see what mischief they could cause . . .

Snatching the list, they looked at
the numbers that came after ten . . .

This is where the pattern begins . . .
Yeh. It is too obvious that all these numbers have 1 ten.
Let's mess them up so kids have no chance of seeing the pattern!
ten
1 ten 1
1 ten 2
1 ten 3
1 ten 4
1 ten 5
1 ten 6
1 ten 7
1 ten 8
1 ten 9
2 tens

The youngest goblin was the first to suggest a terrible idea . . .

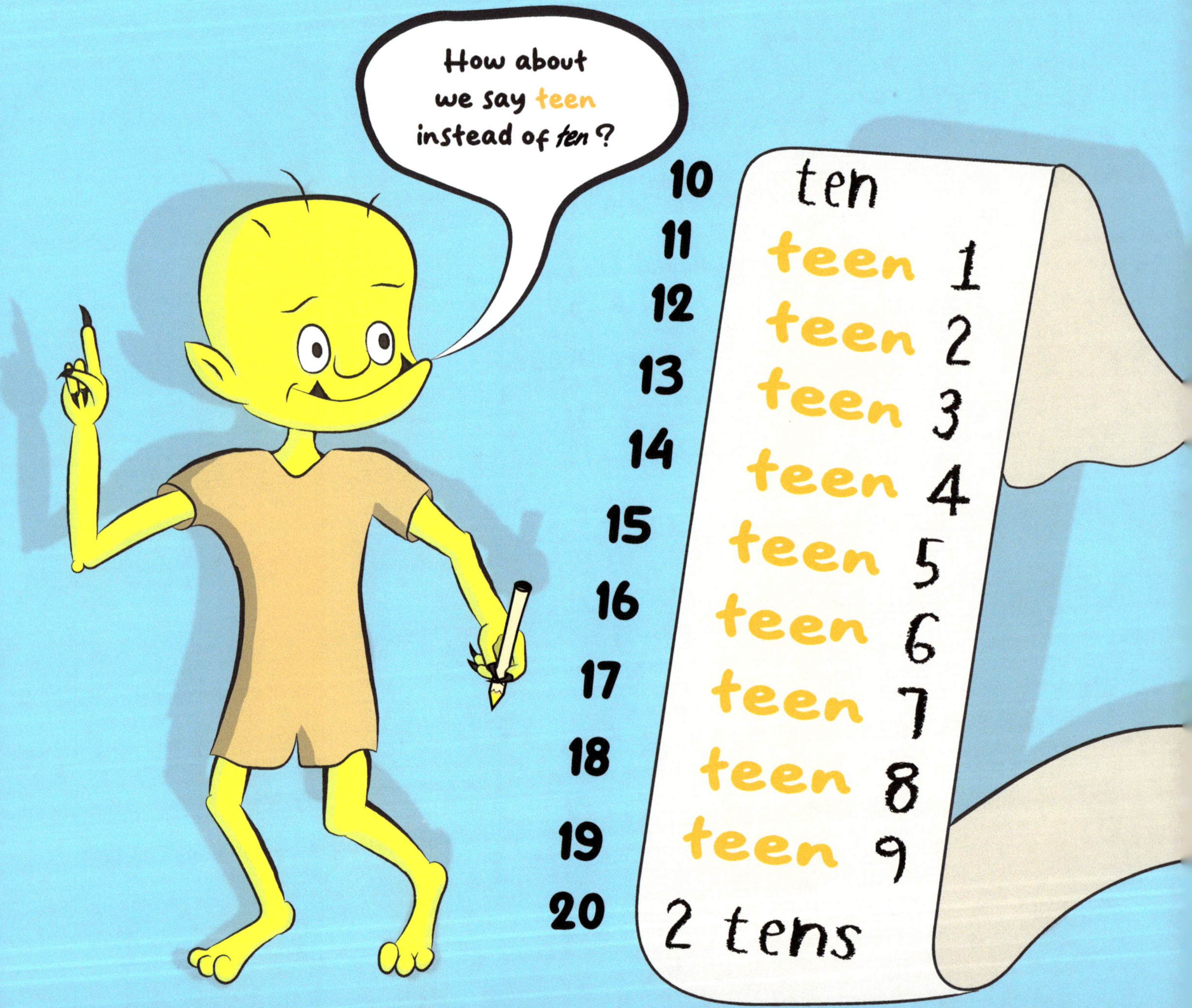

I have an idea. Let's make the teen numbers sound back to front by putting the word teen at the end! Even though teen just means "ten" and the number of tens always comes first.
10 ten
11 1 teen
12 2 teen
13 3 teen
14 4 teen
15 5 teen
16 6 teen
17 7 teen
18 8 teen
19 9 teen
20 2 tens

Let's call the third number after ten thirteen and the fifth one fifteen.
10 ten
11 1 teen
12 2 teen
13 thir teen
4 teen
15 fif teen
16 6 teen
17 7 teen
18 8 teen
19 9 teen
20 2 tens

Now, just in case there is a kid who still isn't confused, I'm gonna do something completely random. I'm going to call these ones eleven and twelve. Ha!
10 ten
11 eleven
12 twelve
13 thir teen
14 4 teen
15 fif teen
16 6 teen
17 7 teen
18 8 teen
19 9 teen
20 2 tens

After completely messing up the numbers after ten, the goblins stood back to admire their mischief. But, they were not finished ...

What's next? Are you going to mess up the tens?
Sounds good to me. I'm sick of saying ten already.
That was my plan. But we must be quick! Master of Order could be back any second...
10 ten
20 2 tens
30 3 tens
40 4 tens
50 5 tens
60 6 tens
70 7 tens
80 8 tens
90 9 tens
100 1 hundred

Saying ten every time takes too long. Let's change the tens to ty.
10 ten
20 2 ty
30 3 ty
40
4 ty
4 ty 1
4 ty 2
4 ty 3
4 ty 4
6 ty
7 ty
80 8 ty
90 9 ty
100 1 hundred

My turn. I'm going to mess with the third ten. Let's call it thirty and let's change the fifth ten to fifty.
Oh, what an evil idea. Now kids are always going to mix up thirty with thirteen and fifty with fifteen. Ha!
ten
2 ty
thirty
4 ty
fifty
6 ty
7 ty
8 ty
9 ty
1 hundred
10
20
30
40
50
70
80
90
100

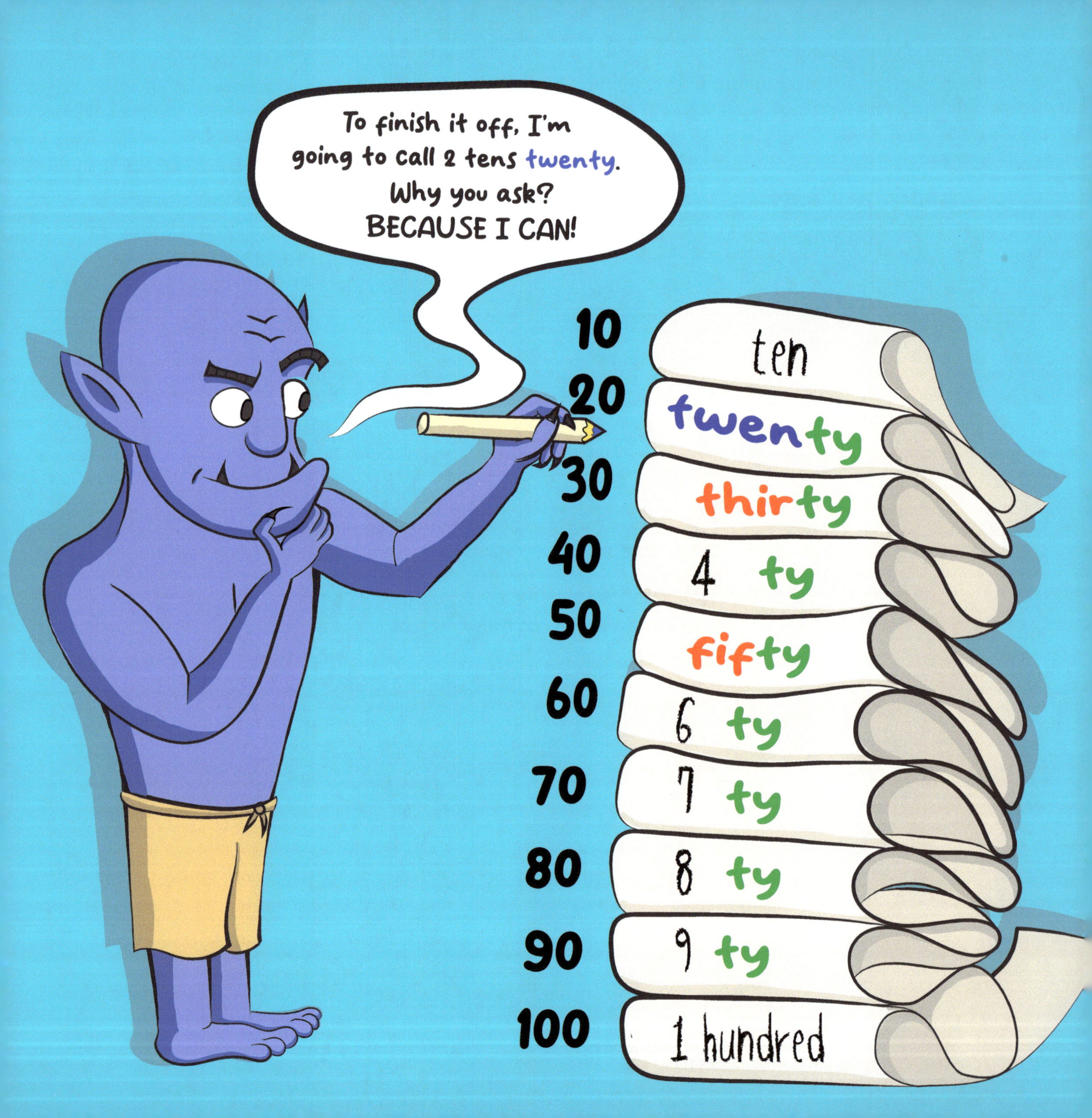
To finish it off, I'm going to call 2 tens twenty. Why you ask? BECAUSE I CAN!
10
ten
20
twenty
30
thirty
40
4 ty
50
fifty
60
6 ty
70
7 ty
80
8 ty
90
9 ty
100
1 hundred

1
2
3
4
5
6
7
8
9
ten
eleven
twelve
thir teen
4 teen
fif teen
6 teen
7 teen
8 teen
9 teen
twenty
twenty 1
twenty 2
twenty 3
twenty 4
twenty 5
twenty 6
twenty 7
twenty 8
twenty 9
thirty
thirty 1
thirty 2
thirty 3
thirty 4
thirty 5
thirty 6
thirty 7
thirty 8
thirty 9
4 ty
4 ty 1
4 ty 2
4 ty 3
4 ty 4
4 ty 5
4 ty 6
4 ty 7
4 ty 8
4 ty 9
fifty
fifty 1
fifty 2
fifty 3
fifty 4
fifty 5
fifty 6
fifty 7
fifty 8
fifty 9
6 ty
6 ty 1
6 ty 2
6 ty 3
6 ty 4
6 ty 5
6 ty 6
6 ty 7
6 ty 8
6 ty 9
7 ty
7 ty 1
7 ty 2
7 ty 3
7 ty 4
7 ty 5
7 ty 6
7 ty 7
7 ty 8
7 ty 9
8 ty
8 ty 1
8 ty 2
8 ty 3
8 ty 4
8 ty 5
8 ty 6
8 ty 7
8 ty 8
8 ty 9
1 hundred
What a mess. It's perfect!
Yep. That will do it.

Finally, after making all the changes, the cheeky Chaos Goblins loaded the number names back into the Super Idea Spreader. Laughing, they plugged it back in and turned it on.

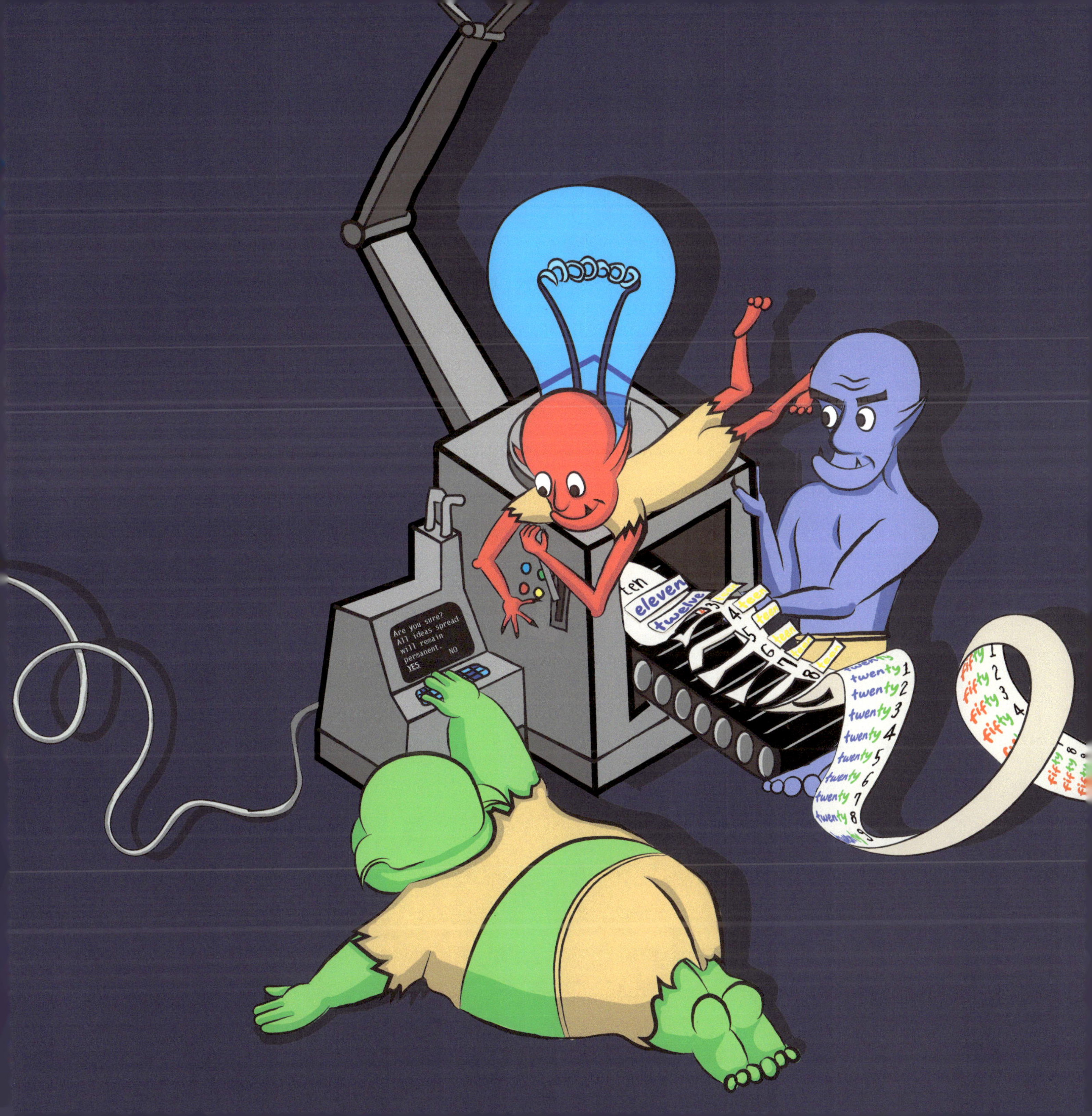

Are you sure?
All ideas spread
will remain
permanent.
YES
NO
ten
eleven
twelve
twenty 1
twenty 2
twenty 3
twenty 4
twenty 5
twenty 6
twenty 7
twenty 8
fifty 1
fifty 2
fifty 3
fifty 4

And, just like that . . .

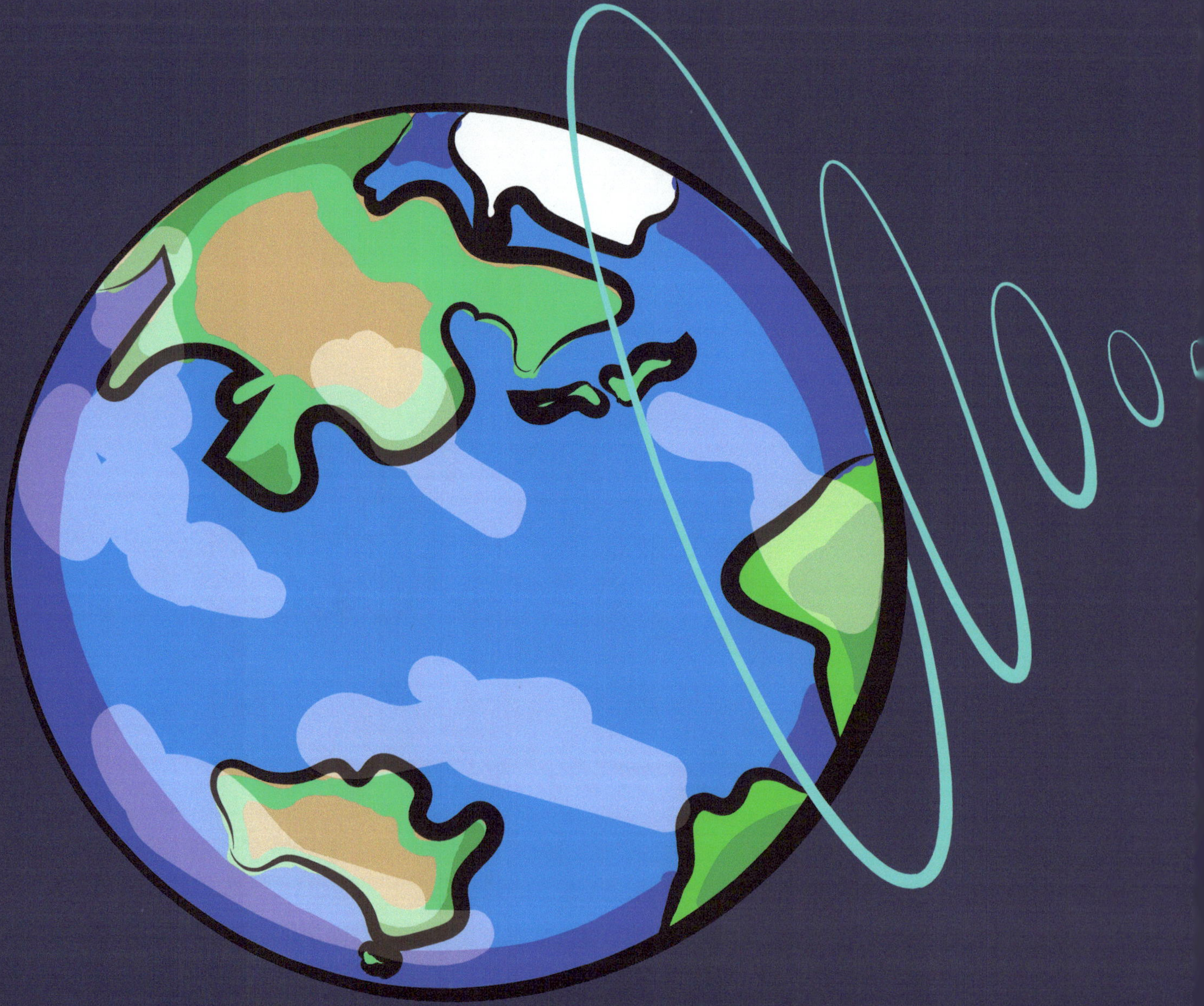

The Super Idea Spreader sent the number names
all over the world.

By the time Master of Order found out what had happened, it was too late. He was furious!

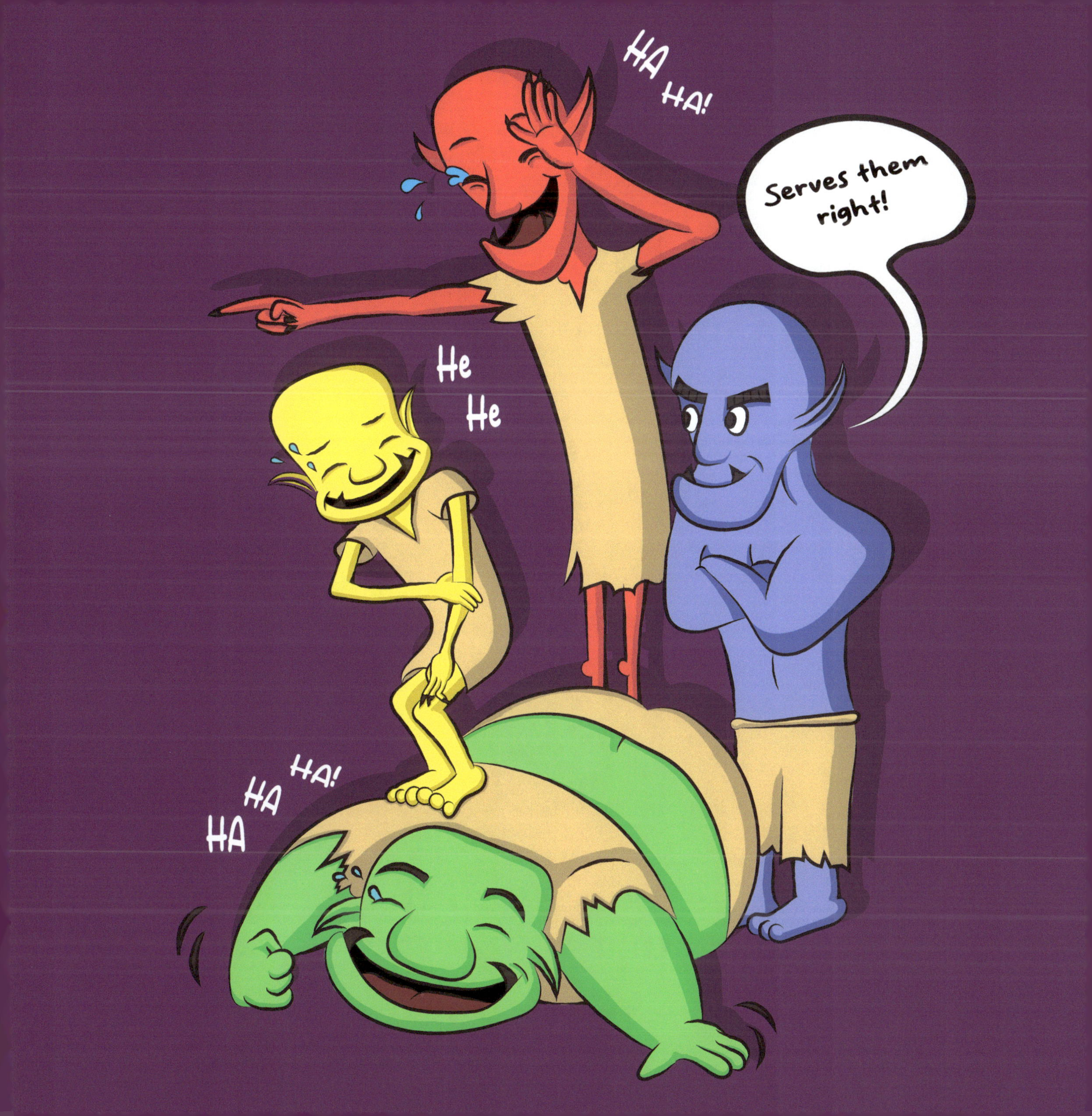
HA
HA!
Serves them right!
He
He
HA
HA
HA!

I wonder . . .
Have you fallen for the goblins' tricks?

or . . .

Can you see the pattern, hiding behind the number names?

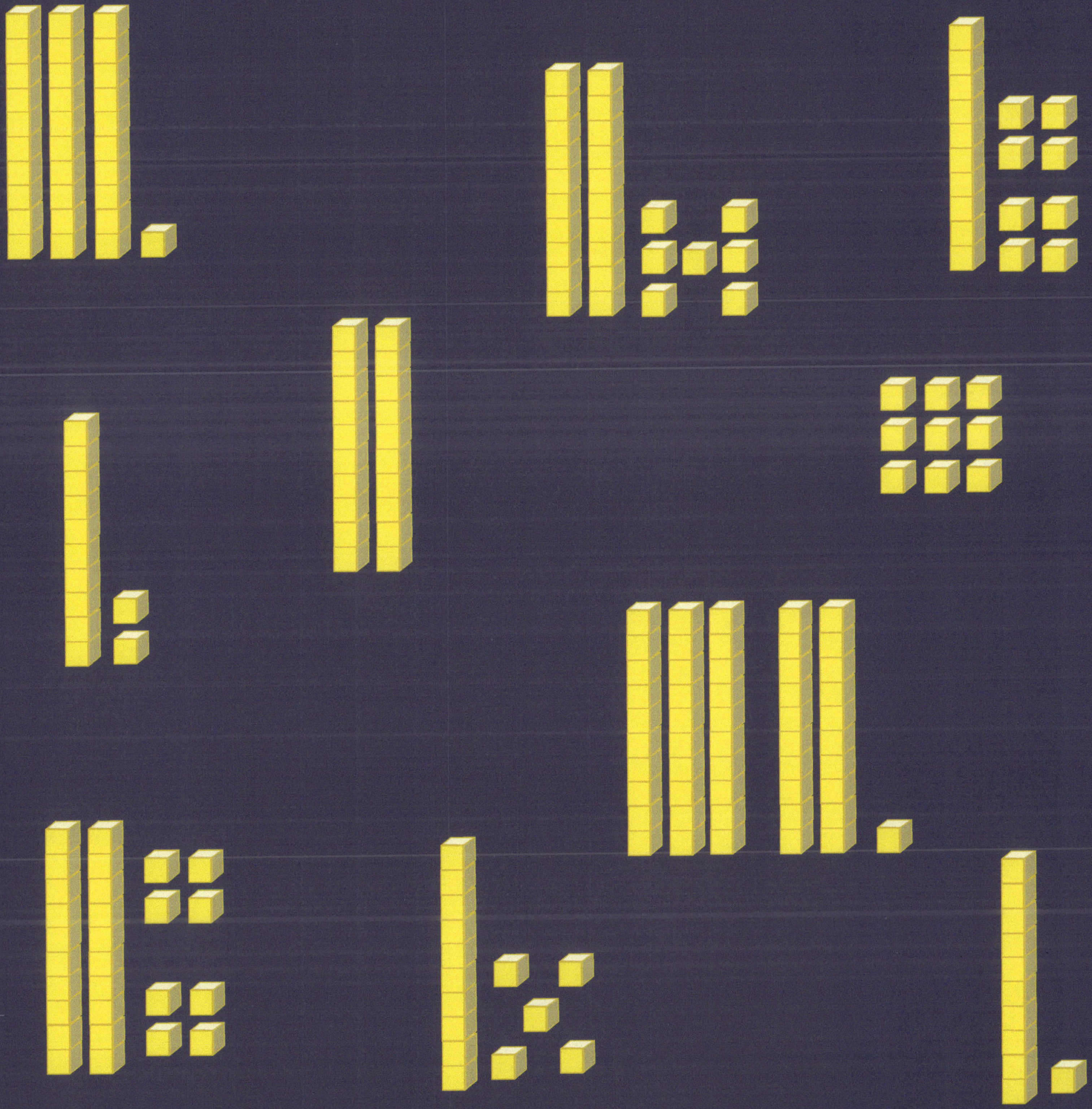

www.ingramcontent.com/pod-product-compliance
Lightning Source LLC
LaVergne TN
LVHW071238210926
848486LV00007B/297

* 9 7 8 1 7 6 4 7 0 7 5 0 3 *